John Torrey

Genealogical Notes,

showing the paternal line of descent from William Torrey, of Combe St.

Nicholas, Somerset County, England, A.D. 1557, to Jason Torrey, of

Bethany, Pennsylvania, with the descendants of Jason Torrey, and of his

brothers and sister

John Torrey

Genealogical Notes,
showing the paternal line of descent from William Torrey, of Combe St. Nicholas,
Somerset County, England, A.D. 1557, to Jason Torrey, of Bethany, Pennsylvania,
with the descendants of Jason Torrey, and of his brothers and sister

ISBN/EAN: 9783337409333

Printed in Europe, USA, Canada, Australia, Japan

Cover: Foto ©berggeist007 / pixelio.de

More available books at **www.hansebooks.com**

WILLIAM TORREY,

OF COMBE ST. NICHOLAS, SOMERSET COUNTY, ENGLAND,

A. D. 1557,

TO

JASON TORREY,

OF BETHANY, PENN'A,

WITH THE

DESCENDANTS OF JASON TORREY,

AND OF HIS BROTHERS AND SISTER,

TO A. D. 1884.

PREFATORY NOTE.

JASON TORREY, popularly known as *Major Torrey*, was one of the earliest settlers in Northeastern Pennsylvania, and was efficiently and conspicuously associated with the first half century of its material and social development.

He raised a large family, and the majority of his descendants have remained in the vicinity of the scenes of his laborious and fruitful activities.

But neither he, nor any of his progenitors, so far as we know, nor any of his descendants, until within a few years, have devoted any attention to the collecting or preserving of any comprehensive genealogical information concerning either the earlier or later generations of the family. Whatever work had been done in this direction, consisted merely of limited family records, made for separate households, and many of these were found to be very incomplete and fragmentary.

In the autumn of 1860, John Torrey, son of Jason Torrey, visited Williamstown, Mass., the place of the latter's birth and early life, and while at the house of

a cousin, who resided on the Homestead, he learned that in the garret there were many old letters which Grandfather had received from his sons after their leaving the parental roof, and had preserved. He arranged with his cousin to select such as were from Jason or his wife, or from his brother Ephraim, who also removed to Bethany, that he might bring them home with him. A large package of letters was thus obtained, written at various dates, from 1793 to the time of the decease of our grandparents.

These letters contained a great amount of interesting and valuable historic information, which it would have been impossible to obtain from any other reliable source.

A careful perusal and re-perusal of these letters, and of Father's early diary, led John Torrey to decide, a few years ago, to endeavor to trace out the ancestry of the family; and he undertook a series of comprehensive and systematic investigations and inquiries, which have been prosecuted with an amount of labor and a degree of expert skill, which no inexperienced person would imagine to be necessary, and which have resulted in bringing to the knowledge of the now living members of the family, a distinct and unbroken line of descent for very near two and a half centuries on this continent; extending back to within twenty years of the landing of the *Mayflower*; and for a hundred years still further

back in England—viz : to the time of a William Torrey, who died at Combe St. Nicholas, in Somerset
County, England, in June, 1557.

Such information as we thus have, respecting those
earlier ancestors in England, was procured by H.
A. Newton, Esq., of Weymouth, Mass., and by him
kindly furnished.

These inquiries have also revealed, incidentally,
many items of interesting and gratifying information
concerning persons of excellent worth, and some of
broad and honorable distinction, who were descended
from the first *William Torrey*, who settled in Weymouth, but outside the line which leads to Jason
Torrey and his descendants.

In collecting the materials for a record of the
families of the descendants of our Williamstown
ancestors, much valuable aid has been given by
M. Homer Torrey, of that town, and by the various other members of those families, of whom in-
quiry has been made

So much of the results of these inquiries as per-
tains to a mere statistical or tabulated Genealogy of
the descendants of Major Jason Torrey, down to the
present date, and of the immediate families of his
father and brothers and sister, and of the direct line
of his paternal ancestry, back to the middle of the
sixteenth century, will be found in the following
pages.

EXPLANATION.

FAMILIES are numbered consecutively. The figures in brackets, at the right of the name of the head of a family, indicates the generation to which such person belongs, counting from William Torrey, who died in 1557.

The figures at the left of a child's name indicate the number under which the *family* of such child will be found.

FAMILIES
OF THE TORREY ANCESTRY.

1

WILLIAM TORREY [1], of Combe St. Nicholas,
in the County of Somerset, England, who died in
June, 1557, leaving a will, in which he mentions
Thomasyne, his wife, and "every of his children,"
without naming them.

2

PHILIP TORREY [2], son of the above William,
and Margaret, his wife. His will is dated in 1604;
mentions his son WILLIAM, and daughter Dorothie:
also his wife Margaret.

3

WILLIAM TORREY [3], son of Philip (No. 2),
and Jane, his wife. His wife died in April, 1639,
at which date he was still living. The date of his
death is not found. His son PHILIP had previously
died.

4

PHILIP TORREY [4], son of William (No. 3), and
Alicie, his wife. He died in June, 1621, leaving a
will dated 21 April, 1621, in which he mentions
three daughters, ANNE, MARY and SARAH, and four
sons, WILLIAM, JAMES, PHILIP and JOSEPH.

The will of the wife Alicie is dated in 1634, and mentions by name the same seven children, and states that the daughter Mary was deceased.

These four sons all emigrated to America in 1640, and located within a few miles of Boston, Mass., and seem to have been the ancestors of all the early families of the name in America. William and Joseph first located in Weymouth ; James in Scituate, and Philip in Roxbury. Although Philip raised a family, it is not learned that he had any *son* to transmit his family name to later generations. He died in Roxbury in 1686.

Joseph removed to Rehoboth, Mass., and subsequently to Newport, R. I., where he was prominently active in public affairs. He died there in 1675. Nothing is learned concerning his descendants. James was accidentally killed by an explosion of powder in Scituate, in July, 1664, leaving five sons and five daughters.

5

WILLIAM TORREY [5], of Weymouth, Mass., son of Philip (No. 4), usually designated as "Captain William Torrey," was born in Combe St. Nicholas, in England, in 1608. The church records show that he was baptized on 21 December of that year.

On 17th March, 1629, he was married to Agnes Combe, of Combe St. Nicholas. She died before he left England, and he was appointed administrator of her estate. In 1640 he came to America, bringing with him two sons, SAMUEL, born in 1632, and WILLIAM, born in 1638, and settled in Wey-

mouth, which was thereafter his home. It is be-
lieved that he brought his second wife with him
from England. Six children were born to him in
Weymouth, viz.: NAOMI, MARY, MICAJAH, JOSIAH,
JUDITH and ANGEL.

Besides which, he took two small children of
his brother James, after their father's death, and
brought them up with his family, viz.: JONATHAN
and MARY. Both of whom are in some of the
records of early history *erroneously* referred to as
his own children.

He died in Weymouth, June 10, 1690. He
was our *earliest* ancestor who resided in America.

SAMUEL TORREY, his eldest son, born in England in 1632, came to New England when but
eight years old ; was educated at Harvard, and
for more than forty years was pastor of the Con-
gregational Church at Weymouth, where he died
in 1707.

6

WILLIAM TORREY [6], of Weymouth, son of
Captain William (No. 5), was born in England in
1608 ; came with his father to Weymouth in 1640;
married Deborah, daughter of John Green ; died
January 11, 1718 ; she died February 8, 1729.
They had eight children, viz.: WILLIAM, JOHN,
SAMUEL, JOSEPH, PHILIP, HAVILAND, JOSIAH and
JANE. His second son, John, was our ancestor
of that generation.

7

JOHN TORREY [7], of Weymouth, (known on
the records as Lieutenant John Torrey), son of
William (No. 6), born 23 June, 1673 ; married

Mary, daughter of Captain William Symes, December 26, 1700; died January 7, 1730; she died February 14, 1758, aged 81. They had seven children, viz., WILLIAM, JOHN, SAMUEL, JOSIAH, MARY, ANN and RUTH. The third son, Samuel, was our ancestor of that generation.

8

SAMUEL TORREY [8], son of John (No. 7), born in Weymouth, February 11, 1706; married Abigail, daughter of William Snowden, of Boston, October 7, 1726. They had six children born in Boston, viz:

ABIGAIL, who married Timothy Scranton.
MARY, who married Ebenezer Hall.
SAMUEL, married and raised a large family.
DAVID, died in childhood.
SARAH, who married Samuel Wilkinson, of Durham.
ANN, who married Hazael Hinman, and after his decease married 2nd, —— Moffatt, of Mass.

In 1736 they removed to Middletown, Conn., where his wife Abigail died, July 12, 1737. He married 2nd, Martha, daughter of Jonah and Martha Strickland, of Middletown, February 16, 1738. They had four children who lived to maturity, viz: JOHN, JOSIAH, WILLIAM, and MARTHA.

In the Spring of 1745, he went as a commissioned company officer, in the Connecticut Regiment of the military expedition to capture Louisburg on Cape Breton Island, and after the capture, he was continued there as one of the officers of the force left as a garrison. He died at Louisburg on 17th September, 1745.

His son Josiah removed from Middletown when a young man, and his subsequent locality is not learned. The daughter Martha became the wife of Abijah Savage, of Middletown, and the mother of a large family. She died 4th June, 1812, aged 67.

9

JOHN TORREY [9], of Williamstown, Mass., eldest son of Samuel of Middletown, Conn., (No. 8), by his second wife Martha, born in Middletown, Conn., January 14, 1741. Removed to Williamstown, Mass., in or about 1766, and secured land for a farm. He married Ruth Tyrell, of Milford, Conn., in 1768, and they resided on their Williamstown homestead and there raised a large family of children. In 1804 he sold his homestead and removed to Richfield, N. Y., where he died in 1823, aged 82. His wife died February 4, 1819.

Their children were:

RUTH, born December 28, 1769; married Stephen Frazer.

NAOMI, born May 1, 1771; married Parsons Ranger.

ASENATH, born October 1, 1772; married Walter Converse; died September 25, 1869; he died September 4, 1852.

JOHN, born December 11, 1774; married Thurza Barnes.

MARTHA, born July 15, 1776; married Abraham Ranger.

JOSEPH, born February 14, 1778; married Mary Giles.

WILLIAM S, born September 27, 1779; died in Infancy.

Anna, born June 4, 1781; married William Hitchings.

Ezra, born January 28, 1784; married Cyntha Briggs.

Grace, born September 23, 1785; married John Ranger.

Abigail, born March 10, 1788; died in Infancy.

Sarah, born July 25, 1792; died young.

10

WILLIAM TORREY [9], of Williamstown, Mass., youngest son of Samuel, of Middletown, (No. 8), by his second wife Martha, born in Middletown, January 21, 1744.

Soon after completing his apprenticeship to a mechanical trade, he in 1766 accompanied his older brother John, to Williamstown, Mass., and secured land for a farm adjoining his brother's, and resided upon it as long as he lived. His mother Martha also removed to Williamstown, and during the last few years of her life, resided alternately with her two sons. She died between 1800 and 1804.

On April 13, 1771, he married Hannah Wheeler, daughter of Dea. Nathan Wheeler, of Williamstown. He died October 30, 1820. She died January 28, 1824.

Their children:

11 Jason, born June 30, 1772.
42 David, born May 19, 1774.
51 Josiah, born March 9, 1776.
52 Samuel, born December 26, 1777.
62 Mary, born May 11, 1780.
72 Ephriam, born March 26, 1782.

11

JASON TORREY [10], of Bethany, Pa., eldest son of William Torrey (No. 10), born in Williamstown, Mass., June 30, 1772. He removed to Mt. Pleasant, Pa., in 1793, and on January 11, 1797, married Lois Welch, daughter of Nathaniel and Lois (Mallery) Welch, then residents of New Ashford, Mass. In February, 1798, he brought her to his Mt. Pleasant home, and in the Autumn of 1801, removed to Bethany. She died July 27, 1813. He married 2nd, Mrs. Achsah Griswold, of Harford, Pa., August 4, 1816. She died Aug. 11, 1830. He died November 21, 1848.

Children by first marriage:

12 WILLIAM, born September 8, 1798.

EPHRIAM, born October 1, 1799; died March 22, 1824.

NATHANIEL, born November 16, 1800; died August 17, 1811.

13 MINERVA, born September 19, 1804.

14 MARIA, born January 1, 1806.

15 JOHN, born April 13, 1807.

16 STEPHEN, born November 9, 1808.

17 ASA, born October 13, 1810.

18 CHARLES, born July 17, 1812.

Children by second marriage:

JAMES, born September 9, 1817; died July 30, 1833.

19 DAVID, born November 13, 1818.

12

REV. WILLIAM TORREY [11], eldest son of Jason (No. 11), born in Mt. Pleasant, Pa., September 8, 1798; studied for the ministry, and on completing his studies went in the Autumn of 1826 to Buenos Ayres, South America, under directions of the Presbyterian General Assembly. In February, 1834, married Miss Elizabeth Sutton, of Buenos Ayres, formerly of Portland, Maine. He returned to the United States in 1836, and died on his farm in Ralls County, Missouri, July 1, 1858.

Children:

20 MARIA ISABEL, born November 23, 1834, in Buenos Ayres.
21 LUCY SUTTON, born January 27, 1838.
22 EMELINE DAVIDSON, born January 4, 1840.
   JASON, born June 29, 1842; died June 30, 1880.
23 WILLIAM JR., born August 22, 1844.
   DAVID, born September 10, 1847; died September 27, 1864.

13

MINERVA TORREY [11], daughter of Jason (No. 11), born in Bethany, Pa., September 19, 1804; married August 7, 1821, Elijah Weston; died November 21, 1875; he died April 26, 1879.

Children:

24 EDWARD W., born December 5, 1823.
25 CHARLES T., born November 28, 1826.
   SARAH P., born February 20, 1829; died August, 1833
   FRANCES E., born July 8, 1832; died April 3, 1857.
   SIDNEY B., born November 1, 1836; died January 12, 1858.
26 MARIA S., born August 28, 1839.
27 MARY A., born December 13, 1843.

MARIA TORREY [11], daughter of Jason Torrey (No. 11), born in Bethany, January 1, 1806; married August 17, 1828, Richard L. Seely; died April 21, 1861 ; he died December 8, 1863.

Their children were:

WILLIAM T. SEELY, born June 24, 1829; died August 23, 1830.

CAROLINE E. SEELY, born March 19, 1831; died March 29, 1832.

WARREN W. SEELY, born May 31, 1832; died August 2, 1845.

28 FRANKLIN A. SEELY, born April 4, 1834.

29 HENRY M. SEELY, born September 18, 1835.

GEORGE D. SEELY, born May 13, 1838.

JOHN T. SEELY, born October 9, 1841; died September 3, 1842.

RICHARD LEWIS SEELY, JR., born October 15, 1842; died November 9, 1847.

15

JOHN TORREY [11], of Honesdale, Pa., son of Jason Torrey (No. 11), born in Bethany, April 13, 1807; married Rebecca Fuller, of Salisbury, Conn., September 28, 1830; she died September 16, 1877.

Children, all born in Honesdale, Pa.:

30 EDWIN F., born June 4, 1832.

ELLEN C., born May 28, 1834; died August 23, 1835.

CAROLINE N., born April 12, 1836.

31 ROBERT N., born August 1, 1838.

32 ADALINE N., born August 15, 1840,

33 HENRY F,, born October 20, 1842.

34 THOMAS F., born October 28, 1844.

35 JOHN, JR., born October 31, 1846.

36 FRANCES REBECCA, born January 18, 1850.

STEPHEN TORREY [11], of Honesdale, son of
Jason (No. 11), born in Bethany, November 9,
1808; married, September 18, 1833, Mary Chapman, of Durham, N. Y.; she died August 4, 1870.

Their children:

JASON, born June 2, 1835; died February 11,
1838.
JAMES, born August 10, 1836; died August 29,
1836.
JASON, born May 31, 1839; married Belle Lewis,
June 15, 1865; died July 20, 1868 ; no child.

ASA TORREY [11], of Bethany, son of Jason
(No. 11), born October 13, 1810; married Polly
G. Brush, February 15, 1832; died December 11,
1884; she died August 31, 1881.

Their children:

37   CHARLES W., born November 26, 1832.
JAMES B., born August 18, 1836; died February
28, 1871.

CHARLES TORREY [11], of Bethany, son of
Jason (No. 11), born July 17, 1812; married
Elizabeth Bruin, of Chatham, New Jersey, April
30, 1834; she died January 14, 1844. Married,
2nd, Mrs. Nancy Kingsbury, of Harford, Pa., October 18, 1844; he died August 14, 1858.

Children by first marriage:

WILLIAM BRUIN, born March 30, 1835; died
March 21, 1848.
ELLEN LUDLOW, born September 3, 1856.

FRANCIS SEELY, born April 11, 1838; died April 20, 1838.

HENRY FULLER, born December 18, 1839; died October 12, 1841.

38   ELIZABETH JANE, born November 4, 1841.

DAVID B., born March 13, 1843; died——1864.

Children by second marriage:

39   ANNA MARIA, born June 24, 1848.

FRANK SEELY, born August 8, 1850; died June 25, 1876.

19

REV. DAVID TORREY, D. D. [11], born in Bethany, Pa., November 13, 1818; married October 3, 1848, Miss Mary E. Humphrey, daughter of Rev. Heman Humphrey, D. D., President of Amherst College, Mass.; she died April 8, 1867. Married, 2nd, Mrs. Georgiana Moseley, of Cazenovia, N. Y., August 28, 1872.

Children by first marriage:

40   SARAH MELISSA, born August 6, 1849.
41   JAMES HUMPHREY, born June 16, 1851.

20

MARIA ISABEL TORREY [12], daughter of Rev. William Torrey (No. 12), born in Buenos Ayres, S. A., November 23, 1834; married March 20, 1861, Charles W. Cleveland.

Children:

NEWCOMB CLEVELAND, born October 7, 1865.
MARY CLEVELAND, born February 10, 1869.

21

LUCY S. TORREY [12], daughter of Rev. William Torrey (No. 12), born in Bethany, Pa., January

27, 1838; married William Cranston, in January, 1860; resides in Ralls County, Missouri.

Children:

JOHN BEDFORD CRANSTON, born October 26, 1860.
DAVID T. CRANSTON, born July 4, 1862.
GRACE CRANSTON, born November 7, 1863.
WILLIAM R. CRANSTON, born February 15, 1866.
CHARLES S. CRANSTON, born February 25, 1870.
ARTHUR CRANSTON, born April 2, 1873.
LENA BELLE CRANSTON, born March 1, 1876.
CLYDE CRANSTON, born September 11, 1878.
LUCY ELIZABETH CRANSTON, born January 22, 1881.

22

EMELINE D. TORREY [12], daughter of Rev. William (No. 12), born in Sparta, N. J., January 4, 1840; married John W. Bedford, of Schuylkill County, Pa., October 20, 1857.

Children:

ALONZO F. BEDFORD, born August 7, 1863.
JOHN CLAUDE BEDFORD, born May 28, 1866.
THEODORE TORREY BEDFORD, born November 12, 1868.
LUCY BELLE BEDFORD, born April 1, 1871; died December 13, 1876.
EDITH BEDFORD, born July 17, 1873; died May 16, 1876.

23

WILLIAM TORREY, JR. [12], son of Rev. William (No. 12), born in Sparta, N. J., August 22, 1844; married Jane Gorrell, of Schuylkill County, Pa., June, 1865.

Children:

FRANK, born March 26, 1866.

MARY ANN, born May 25, 1868; died January 18, 1879.

FANNY SUTTON, born June 17, 1870.

ROBERT GORRELL, born May 17, 1872.

JENNIE BELLE, born March 10, 1876.

24

EDWARD W. WESTON [12], of Scranton, Pa., son of Minerva T. Weston and Elijah Weston (No. 13), born December 5, 1823; married Susan S. Moore, of Honesdale, June 4, 1852.

Children:

CAROLINE S. WESTON, born April 11, 1853.

CHARLES S. WESTON, born April 25, 1860.

ROBERT M. WESTON, born June 20, 1869; died December 7, 1869.

25

CHARLES T. WESTON [12], son of Minerva T. and Elijah Weston (No. 13), born November 28, 1826; married Mary C. Crawford, of Woodstock, Va., June 10, 1852; she died October 4, 1864. Married, 2nd, Marcia E. Lent, of Scranton, Pa., June 6, 1869; he died July 7, 1884.

Children by first marriage:

WILLIAM E. WESTON, born April 12, 1853; died March 10, 1856.

EDWARD C. WESTON, born February 8, 1856; died November 26, 1869.

LENA MAY WESTON, born December 28, 1858; died May 10, 1863.

ELEANOR M. WESTON, born July 4, 1862.

JAMES WESTON, born April 6, 1864; died September 25, 1864.

Child by second marriage:

LOUIS LENT WESTON, born May 31, 1870: died July 30, 1870.

26

MARIA S. WESTON [12], daughter of Minerva T. and Elijah Weston (No. 13), born August 28, 1839; married James H. Bartlette, of Virginia, February 9, 1860; he died November 23. 1879.

Children:

HENRY C. BARTLETTE, born November 22, 1860.
CHARLES W. BARTLETTE,born April 13, 1862.
MARY G. BARTLETTE, born August 5, 1864: died January 14, 1865.
LUCY M. BARTLETTE, born August 8, 1866.
JOSEPH C. BARTLETTE, born July 8, 1868; died August 5, 1868.
EDWARD W. BARTLETTE, } born January 30, 1871.
JAMES H. BARTLETTE.   }

27

MARY A. WESTON [12], daughter of Minerva T. and Elijah Weston (No. 13), born December 13, 1843; married James W. Belknap, February 28, 1870; resides in Topeka, Kansas.

Have no children.

28

FRANKLIN A. SEELY [12], son of Maria T. and Richard L. Seely (No. 14), born at Seelys Mills, Pa., April 4, 1834: married Mary G. Wessells, November 17, 1858; she died July 13, 1876; he resides in Washington, D. C.

Children:

LEWIS WARREN SEELY, born December 10, 1859.
HENRY W. SEELY, born November 15, 1861.

MARY SEELY, born January 9, 1863.

FREDERIC G. SEELY, born December 31, 1863; died July 19, 1864.

FRANK HOWARD SEELY, born April 18, 1868.

FLORENCE HELENA SEELY, born September 15, 1874; adopted by her uncle Henry M. Seely.

29

HENRY M. SEELY [12], son of Maria T. and Richard L. Seely (No. 14), born at Seelys Mills, Pa., September 18, 1835; married, May 13, 1862, Miss Kate Tracy, of Honesdale. They reside in Honesdale. They adopted the youngest child of Franklin A. Seely (No. 28), FLORENCE H. SEELY, born September 15, 1874; died May 3, 1884.

30

EDWIN F. TORREY [12], of Honesdale, Pa., son of John Torrey (No. 15), born in Honesdale, June 4, 1832; married October 5, 1854, Jennette S. Stone, of Honesdale.

Children:

GEORGE NIVEN, born November 11, 1856; died November 28, 1860.

JOHN HENRY, born September 11, 1860.

WILLIAM STONE, born July 12, 1862.

CATHARINE REBECCA, born June 6, 1866.

EDWIN FULLER, JR., born November 11, 1870.

31

ROBERT N. TORREY [12], of Honesdale, son of John Torrey (No. 15), born in Honesdale, August 1, 1838; married, February 23, 1864, Elizabeth D. Thompson, of Port Jervis, N. Y.

Children:

Augustus Thompson, born May, 1865; died two days after.

Clara Rebecca, born July 18, 1866.

Ada Grant, born June 24, 1871; died October 3, 1878.

Edith Fuller, born June 3, 1880.

32

ADALINE N. TORREY [12], daughter of John Torrey (No. 15), born in Honesdale, Pa., August 15, 1840; married Dr. James A. S. Grant, of Cairo, Egypt, February 7, 1870. Resides in Cairo, Egypt.

Children:

William Torrey Grant, born January 20, 1871.

Jessie Campbell Grant, born August 10, 1873.

33

HENRY F. TORREY [12], of Montclair, N. J., son of John Torrey (No. 15), born in Honesdale, Pa., October 20, 1842; married, November 23, 1869, Harriet A. Pratt, of Montclair, N. J.

Children:

Julius Pratt born February 8, 1871.

Henry Fuller, born September 30, 1872; died July 22, 1874.

James Eaton, born July 2, 1876.

Robert Grant, born July 12, 1878.

Gertrude, born July 2, 1880.

Arthur Morris, born December 30, 1882.

THOMAS F. TORREY [12], of Scranton, son of John Torrey (No. 15), born in Honesdale, Pa., October 28, 1844; married, February 16, 1869, Sophia R. Dickson, daughter of Thomas Dickson, of Scranton, Pa.

Child:

THOMAS DICKSON TORREY, born September 15, 1881.

35

JOHN TORREY, JR. [12], of California, son of John Torrey (No. 15), born in Honesdale, Pa., October 31, 1846; married, December 31, 1874, Emma Tabor, of San Francisco, Cal.

Children:

LOUISA ALICE, born November 3, 1875.
JOHN FRANCIS, born July 16, 1879.

36

FRANCES R. TORREY [12], daughter of John Torrey (No. 15), born in Honesdale, January 18, 1850; married, January 29, 1873, Andrew Thompson, of Port Jervis, now of Honesdale.

Children:

AUGUSTUS P. THOMPSON, born December 29, 1873.

ROBERT TORREY THOMPSON, born September 8, 1876; died October 8, 1877.

REBECCA FULLER THOMPSON, born March 22, 1881.

37

CHARLES W. TORREY [12], of Bethany, son of Asa (No. 17), born in Bethany, November 26,

1832; married Elizabeth G. Graves, of Bethany, July 1, 1855; died October 10, 1882.

Only child:

LILIAN E. TORREY, born August 17, 1879; married Charles W. Babbitt, of Honesdale, May 1, 1878. Children—*Ida T. Babbitt*, born January 20, 1879; *Lizzie G. Babbitt*, born December 5, 1881.

38

ELIZABETH JANE TORREY [12], daughter of Charles Torrey (No. 18), born in Bethany, Pa., November 4, 1841; married, September 14, 1865, Col. Frederick H. Harris, of Montclair, N. J.

Children:

ELLEN TORREY HARRIS, born September 14, 1866.

ELIZABETH T. HARRIS, born July 25, 1870.
JANE HOWELL HARRIS, born July 11, 1872.
FREDERICK HALSEY HARRIS, born August 3, 1877; died March 27, 1879.
ANNA M. HARRIS, born November 10, 1879.

39

ANNA MARIA TORREY [12], daughter of Charles Torrey (No. 18), born in Bethany, Pa., January 24, 1848; married, September 10, 1867, John C. Cheney, of Fort Dodge, Iowa.

Children:

JOHN T. CHENEY, born May 28, 1868.
CARL FRANK CHENEY, born October 29, 1872.

40

SARAH MELISSA TORREY [12], daughter of Rev. David Torrey (No. 19), born in Honesdale,

Pa., August 6, 1849: married William Delos
Wells, of Cazenovia, N. Y., September 24, 1873.
Resides in Cazenovia.

Children:

MARY ADALAIDE WELLS, born July 16, 1874.
SARAH HUMPHREY WELLS, born December 19,
1875.
DAVID TORREY WELLS, born May 30, 1882.

41

JAMES H. TORREY, [12], of Scranton, son of
Rev. David Torrey (No. 19), born in Delhi, N.
Y., June 16, 1851; married, December 10, 1872,
Ellen C. Jay, of Scranton.

Children:

MARY H. TORREY, born January 31, 1874.
WILLIAM J. TORREY, born June 24, 1875.
ELIZABETH J. TORREY, born March 18, 1878.

Descendants of David Torrey.

42

DAVID TORREY [10], of Williamstown, Mass.,
son of William (No. 10), born in Williamstown,
May 19, 1774; married Esther Woodcock, January 7, 1799; died December 17, 1853; she died
November 25, 1852.

Children:

43 WILLIAM, born October 13, 1799.
    Lois, born August 8, 1801; died August 26, 1843, unmarried.
44 MYRON, born April 27, 1803.
45 ANNA, born July 3, 1805.
46 NORMAN, born June 24, 1807.
    DAVID CICERO, born September 27, 1809; died November 15, 1832, unmarried.
47. JUSTIN, born April 25, 1812.
48 ESTHER, born September 5, 1814.
49 CORDELIA, born May 9, 1818.
50 JASON E., born November 13, 1820.

43

WILLIAM TORREY [11], of Williamstown, son of David (No. 42), born October 13, 1799; married Lydia Newton, March 26, 1824; died April 10, 1881; she died ———— 24, 1877.

Having no child, they adopted a daughter of *his* sister and her brother, named:

MINERVA NEWTON, born July 29, 1840; married Lewis J. Gardner, September 19, 1877; died April 20, 1882, leaving two sons—*William T. Gardner*, born December 26, 1879, and *Henry N. Gardner*, born April 11, 1882.

44

MYRON TORREY [11], of Williamstown, son of David (No. 42), born April 27, 1803; married Harriet Converse, September 17, 1834; died December 3, 1879.

Children:

OLIVE A. TORREY, born November 8, 1835; married Jeremiah J. Pratt, December 14, 1861;

children—*Emma J. Pratt*, born August 23, 1862; married William H. Kelly, September 7, 1880; (children—H. Garfield Kelly, born September 1, 1881; William Henry Kelly, born June 18, 1884.) *Grace L. Pratt*, born July 5, 1865; *Torrey Pratt*, born August 7, 1872; died October 7, 1872; *Warner Pratt*, born June 15, 1874; died April 23, 1877; *Carl W. Pratt*, born June 20, 1877.

EMMA E. TORREY, born February 26, 1837; died August 31, 1844.

D. WALLACE TORREY, born May 16, 1839; married Mary E. Seaver, December ——, 1869; she died November 7, 1880.

JESSIE A. TORREY, born November 5, 1841; married William Drew, September 17, 1868; children—*Howard Drew*, born December 30, 1869; *Harriet B. Drew*, born September 28, 1872; *Robert Drew*, born June 13, 1875.

H. ABBY TORREY, born June 4, 1846; died December 25, 1872.

M. HOMER TORREY, born November 8, 1850.

CLARIBEL TORREY, born February 24, 1853; married Monroe C. Loveland, September 28, 1880; died November 16, 1884.

45

ANNA TORREY [11], daughter of David (No. 42), born July 3, 1805; married Ezra Newton, October 16, 1825; died May 24, 1842.

Children :

D. TORREY NEWTON, born ——, 1826; married Phebe Ogden, November 22, 1848; children—*Nellie A. Newton*, born December 4, 1861; married Milton Anderson,——; *Carrie E. Newton*, born January 16, 1871.

LYDIA E. NEWTON, born June 14, 1828; married

Almon A. Barrett, September 8, 1844; child —*Lucy N. Barrett*, born November 9, 1848; married Albert J. Crane, June 21, 1879; (child—Wallace B. Crane, born September 22, 1882.)

GEORGE A. NEWTON, born April 22, 1831; married Charlotte Batton, February 16, 1857; children—*Owen E. Newton*, born December 4, 1858; *Ella E. Newton*, born October 15, 1861; died September 30, 1862.

ESTHER E. NEWTON, born May 24, 1833; married Joel Carpenter, January 14, 1864.

MARTHA A. NEWTON, born December 1, 1835; married Benjamin Hayward, ————; child— *Nellie M. Hayward.*

WILLIAM NEWTON, born April 20, 1838; died in Hospital of Army, in Kentucky, May 31, 1864.

MINERVA NEWTON, born July 29, 1840; adopted by her uncle William Torrey (No. 43), which see.

46

NORMAN TORREY [11], son of David (No. 42), born June 24, 1807; married Ann Krigger, September 22. 1832.

Children:

CAROLINE TORREY, born September 28, 1833; died March 1, 1864.

ELIZA TORREY, born December 12, 1834; died May 12, 1842.

CICERO TORREY, born July 27, 1837; married Viola Anderson, November 1, 1883; child——, born in 1884.

EVEREL TORREY, born March 21, 1839; married Julia A. Hilby, December 5, 1877.

ELIZA A. TORREY, born August 24, 1843; died August 11, 1872.

HOMER TORREY, born December 4, 1846; died February 28, 1848.

JUSTIN TORREY [11], son of David (No. 42),
born April 25, 1812; married Olive Converse,
April 6, 1837; died June 14, 1870.

Children:

J. CICERO TORREY, born February 11, 1838;
married Adaline Hitchcock, November 16, 1862;
children—*Minnie*, born August 11, 1863; *Lucy*,
born March 15, 1865; *Alice*, born January 26,
1870; died January 20, 1875; *Bertha*, born July
31, 1879.
KATHARINE TORREY, born May 12, 1841.
ADOLPHUS TORREY, born December 15, 1844;
died in the Army while a prisoner at Belle Island,
February 23, 1862.
GEORGE B. TORREY, born April 12, 1849.
D. CARLTON TORREY, born December 3, 1852;
died March 12, 1857.
W. CARLTON TORREY, born April 10, 1857; died
March 14, 1858.
MABEL O. TORREY, born August 28, 1859; mar-
ried Frank A. Smith, March 7, 1883.

ESTHER TORREY [11], daughter of David (No.
42), born September 5, 1814; married Webster
Noyes, of Williamstown, October 12, 1848; died
August 25, 1882; he died February 28, 1866.

Children:

SARAH E. NOYES, born October 7, 1849; *drowned*
September 4, 1880.
SAMUEL T. NOYES, born November 26, 1850.
FLORENCE E. NOYES, born February 12, 1853.
CHARLES W. NOYES, born January 13, 1855
MARY O. NOYES, born May 5, 1857.

CORDELIA TORREY [11], daughter of David (No. 42), born May 9, 1818; married William A. Morey, of Williamstown, March 27, 1839.

Children:

FRED A. MOREY, born October 21, ~~1830~~. *1840*

ALMA E. MOREY, born January 23, 1842.

E. ELIZABETH MOREY, born May 20, 1844; married Lester Eldridge, February 22, 1866; died August 25, 1867.

ANNA A. MOREY, born February 28, 1851; died January 26, 1858.

GEORGE W. MOREY, born January 7, 1853; died September 1, 1875.

50

JASON E. TORREY [11], son of David (No. 42), born November 13, 1820; married Mary Lewis, April 17, 1851.

Children:

LEWIS E. TORREY, born March 6, 1852; married Ella Ludden, July 1, 1880.

REUBEN H. TORREY, born February 18, 1854; died May 6, 1882.

EMERSON M. TORREY, born March 25, 1856; married Addie M. Graves, January 1, 1879; children—*Annie L. Torrey*, born March 7, 1880; *Charles L. Torrey*, born April 12, 1881; died May 7, 1881; *Clarence B. Torrey*, born February 21, 1883.

D. CLARENCE TORREY, born January 29, 1859.

ESTHER M. TORREY, born May 3, 1861.

INEZ C. TORREY, born July 6, 1865.

E. DANA TORREY, born May 9, 1869.

JOSIAH TORREY [10], son of William (No. 10), born in Williamstown, March 9, 1776; married Jane McDonald of New York, November 27, 1810; died while a staff officer in the army, stationed on Staten Island, November —, 1814.

Children:

MARY JANE TORREY, born in New York, October 8, 1811; married Samuel Hubbell, of Lanesboro, a widower with a family. She died without issue.

DAVID WHEELER TORREY, born March 9, 1813, died young, but date not learned.

--- • • ---

Descendants of Samuel Torrey.

52

SAMUEL TORREY [10], son of William (No. 10), born in Williamstown, Mass., December 26, 1777; married Elizabeth Lewis, January 10, 1799. She died December 13, 1818; he married 2nd, Mrs. Submit Ellis, of Lisle, N. Y., March 21, 1819; died September —, 1854; she died November 12, 1867.

Children by first marriage:

53 POLLY, born October 9, 1799.
54 DAVID, born November 21, 1800.
55 GEORGE B., born February 7, 1802.
56 AZUBAH G., born November 21, 1803.

Jason, born November 4, 1805; married but had no child; died about 1870.

57   Nathan W., born February 1, 1807.

John Milton, born October 23, 1809; died June 24, 1878, unmarried.

Lucy Ann, born July 6, 1812; died November 13, 1843, unmarried.

Amos, born November 5, 1814; died but date not learned.

Children by second marriage:

58   Elizabeth N., born January 14, 1820.

59   Nancy J., born January 27, 1822.

60   Josephus, born September 22, 1824.

61   Hannah, born May 22, 1828.

53

POLLY TORREY [11], daughter of Samuel (No. 52), born October 9, 1799; married Samuel Walton, of Lisle, New York, June —, 1821; died October 6, 1857; he died September 26, 1846.

Children:

Louisa Walton, born April 1, 1822; died March 15, 1848.

Maria Walton, born November —, 1823; died January 13, 1841.

Almira Walton, born January 18, 1825; died January 2, 1841.

Lewis Walton, born March 14, 1827; married Susan A. Francis, February 17, 1864.

Otis Walton, born October 1, 1829; died July 23, 1853.

Oliver Walton, born March 30, 1833; married in 1864.

Washington Walton, born July 14, 1836; married Francis M. Wright, September 7, 1867;

she died May 16, 1878; married 2nd, Leah Tressler, February 2, 1879. Children by first marriage—*Otis Walton, Jr.*, born June 20, 1868; *Minnie J. Walton*, born August 15, 1870.

54   1593404

DAVID TORREY [11], son of Samuel (No. 52), born November 21, 1800; married Hannah K. Smith, March 7, 1843; she died June —, 1878.

Children:

SILAS TORREY, born March 4, 1844; living unmarried.

ISAAC TORREY, born October 5, 1845; married Dilla Maria Hamilton, October 30, 1872; child —*Dana Torrey*, born October 7, 1879.

ANNA E. TORREY, born June 10, 1847; married George R. Hallenbeck, February 22, 1871; children—*Ray Hallenbeck*, born May 5, 1873; *Lilian M. Hallenbeck*, born December 9, 1874; *Minnie E. Hallenbeck*, born June 1, 1878.

LUCIUS TORREY, born June 6, 1849; married Adie E. Kinsman, November 20, 1878; children— *Nellie May Torrey*, born October 18, 1880; *Gertrude Torrey*, born May 15, 1883.

AMANDA M. TORREY, born October 15, 1854; married Clark B. Hamilton, January 1, 1872.

55

GEORGE B. TORREY [11], son of Samuel (No. 52), born February 7, 1802; married Elizabeth B. Boardman, January 10, 1826; died March 8, 1876.

Son:

EZRA NEWELL TORREY, born November 2, 1828; married Nancy M. Hotchkiss, June 3, 1852; chil-

dren—*William W. Torrey*, born September 29, 1853; married Lydia Wilfong, November 29, 1877; (children—Fidelia May Torrey, born February 9, 1879; Nancy E. Torrey, born April 5, 1881, and Joseph Newell Torrey, born March 20, 1883); *Samuel Torrey*, born June 14, 1856; married Anna M. Steiner, March 6, 1883; *Mary E. Torrey*, born June 14, 1856; married James L. Gregory, October 11, 1874; (children—Vilette Gregory, born—, 1875; Philip Gregory, born —— 1877; Samuel Gregory, born ——, 1879); *Charlotte E. Torrey*, born February 12, 1860; died April 23, 1860; *Lydia M. Torrey*, born February 17, 1869.

56

AZUBAH G. TORREY [11], daughter of Samuel (No. 52), born November 21, 1803, married John Boardman, Jr., November 21, 1826; he died December 10, 1841; married 2nd, —— Foote; she died February 22, 1882.

Children:

JAMES BOARDMAN, born October 22, 1835; married Melissa Castle, November —, 1872; child—*John S. Boardman*, born September 5, 1873.

MARY BOARDMAN, born July 8, 1838; married John H. Clark, August 18, 1861; children—*Myron H. Clark*, born April 23, 1867; *Ida W. Clark*, born September 11, 1868; *J. Merton Clark*, born December 6, 1869; *Henry P. Clark*, born March 11, 1871; *George D. Clark*, born July 25, 1872; *William H. Clark*, born February 11, 1874; died February 9, 1881; *Lewis W. Clark*, born August 6, 1881, and *Leon E. Clark*, born March 19, 1883.

NATHAN W. TORREY [11], son of Samuel (No.
52), born February 1, 1807; married and had
several children, but particulars are not obtained.

ELIZABETH N. TORREY [11], daughter of Samuel (No. 52), born January 14, 1820; married
—————— Delano; died September 11, 1849.

Children:

ELLEN DELANO, born —————— ; married ——
Horton; children—*one son* and *one daughter*;
names not learned; —— Delano, an unmarried
daughter living.

NANCY JUDSON TORREY [11], daughter of
Samuel (No. 52), born January 27, 1822; married Lucius Lindsley in 1845; he died in 1849;
married 2nd, Henry M. Richardson in 1852.

Children:

ERMINA J. LINDSLEY, born June 4, 1846; married Adams Gibson, March 17, 1864; children—
Oscar B. Gibson, born June 9, 1865; *Ida M.
Gibson*, born October 6, 1867; *Clarence W. Gibson*, born November 13, 1869; *Mary E. Gibson*,
born November 19, 1872; *Ellie N. Gibson*, born
August 30, 1874; *Hattie E. Gibson*, born May
18, 1876; *Ralph R. Gibson*, born December 1,
1879; *Blanche M. Gibson*, born December 17,
1881.

—————— LINDSLEY, (second daughter), born in
1848; married ——————; died in Michigan in
1881, leaving *four sons* and *one daughter*; particulars not obtained.

Children by second husband :

FRANK RICHARDSON, born ————; married
————; has *two sons.*

ADRIAN RICHARDSON, born————.

ALFRED RICHARDSON, born ————.

JOHN RICHARDSON, born February 21, 1865.

60

JOSEPHUS TORREY [11], of Johnstown, N. Y.,
son of Samuel (No. 52), born September 22, 1824;
married Nancy M. Cosselman, December 31,
1849; died March 26, 1876.

Children:

CELIA A. TORREY, born November 22, 1850;
married George M. Simmons, December 18, 1869;
children—*Edna T. Simmons,* born April 16, 1876;
Clarence A. Simmons, born September 11, 1879;
Bertha A. Simmons, born February 16, 1881.

EDWARD E. TORREY, born January 13, 1852;
died February 15, 1852.

ALVIN TORREY, MELVIN TORREY and EMMA
TORREY all died in infancy.

MARY E. TORREY, born October 19, 1855; died
March 5, 1863.

FRED S. TORREY, born March 28, 1857.

FRANK E. TORREY, born August 15, 1858; married Emma Calkins, December 16, 1883.

GEORGE TORREY, born November 28, 1869; died
January 21, 1870.

61

HANNAH TORREY [11], daughter of Samuel
(No. 52), born May 22, 1828; married Orator
Gibson, October 26, 1854.

Children:

JESSE GIBSON, born October 26, 1855.
NELLIE E. GIBSON, born August 17, 1857.
EUGENE GIBSON, born October 27, 1859; married ————, October 1, 1880; child—*Leroy Gibson*; born December 12, 1881.
VIOLA GIBSON, born July 24, 1861.
LEWIS GIBSON, born December 25, 1863.
CHARLES GIBSON, born December 28, 1866
LEO GIBSON, born January 30, 1870.
GUY GIBSON, born February 19, 1873.

————•••————

Descendants of Mary Torrey Smith.

62

MARY TORREY [10], daughter of William Torrey (No. 10), born in Williamstown, Mass., May 11, 1780; married Stephen Smith, January 26, 1800; died March 25, 1834; he died May —, 1838.

Children:

HANNAH SMITH, born May 27, 1801; married Nehemiah G. Philos, of Half Moon, N. Y., in 1838; died February 14, 1853 ; no children.
MARILLA SMITH, born October 2, 1802 ; married David Newland, of Stillwater, N. Y.; died September, —, 1877 ; no children.
63   WILLIAM SMITH, born October 27, 1804.
64   JOHN L. SMITH, born June 9, 1807.
65   MARY SMITH, born September 10, 1810.
66   JANE AMELIA SMITH, born January 25, 1813.

67 JOSIAH T. SMITH, born August 4, 1815.
68 MARTHA ANN SMITH, born August 3, 1817.
69 STEPHEN HENRY SMITH, born July 28, 1819.
70 LUCIUS EDWIN SMITH, born January 29, 1822.
71 ELISHA HUBBELL SMITH, born April 13, 1825.

63

WILLIAM SMITH [11], son of Mary Torrey
Smith (No. 62), born October 27, 1804; married
Sarah Phillips, of Adams, Mass., in 1831; died
April 16, 1862.

Children:

JULIA ANN SMITH, born in 1832; married ——
Herring in 1856; died in 1859; no children.
WILLIAM ROSCOE SMITH, born in 1836; married
Sarah Sweet, of Albion, N. Y.; has *one daughter.*

64

REV. JOHN L. SMITH [11], now of Deposit, N.
Y., son of Mary Torrey Smith (No. 62), born
June 9, 1807; married Francis C. Oviatt, May 9,
1839.

Children:

FIVE DAUGHTERS, all of whom are deceasad.

65

MARY SMITH [11], daughter of Mary Torrey
Smith (No. 62], born September 10, 1810; married Edwin W. Phelps, in 1844; died in April,
1883.

Children:

MARY FRANCES PHELPS, deceased; unmarried.
TWO SONS; both deceased.

66

JANE AMELIA SMITH [11], daughter of Mary
Torrey Smith (No. 62), born January 25, 1813;
married Gilbert B. Smith, of Williamstown, Mass.,
in 1840; died September 9, 1855.

Only child:

WATERMAN SMITH, ———; died September 9,
1855.

67

REV. JOSIAH T. SMITH [11], son of Mary Torrey
Smith (No. 62), born August 4, 1815; married
Harriet Amelia Black, August 23, 1844.

Children:

FRANCIS B. SMITH, born March 31, 1846; married————; have *one son* and *one daughter*
living.

MARY SMITH, born March 8, 1848; deceased.

ARTHUR NEAL SMITH, born January 6, 1850;
married; has *one daughter.*

RUTH A. SMITH, born May 15, 1852; died in
infancy.

EDWARD G. SMITH, born March 20, 1855; died
September 28, 1856.

HATTIE J. SMITH, born January 1, 1859; died
January 26, 1868.

JANE A. SMITH, born February 8, 1861; died
March 10, 1868.

68

MARTHA ANN SMITH [11], daughter of Mary
Torrey Smith (No. 62), born August 3, 1817;
married Thomas Ingham, of Sandisfield, Mass., in
1848.

Children:

MARY FRANCIS INGHAM, born ———; married
Edwin James, of Lewisburg, Pa.; have *two sons*
and *one daughter.*

LUCIUS INGHAM.   No particulars obtained.

EDWARD INGHAM.   No particulars obtained.

69
DR. STEPHEN H. SMITH [11], son of Mary
Torrey Smith (No. 62), born July 28, 1819; mar-
ried Martha L. Adair, May 26, 1850.

Children:

LUCIUS A. SMITH, born June 18, 1852; died
September 3, 1852.

FRANCES JANE SMITH, born January 15, 1856;
married Elijah A. Hackworth, January 26, 1880;
children—*Eleanor Louisa Hackworth*, born December 4, 1881; *Blanche Adair Hackworth*, born
June 27, 1883.

JOHN HENRY SMITH, born August 4, 1859; married ———; died September 3, 1884; no children.

MARGARET LOUISA SMITH, born March 20, 1864.

IDA NEWLAND SMITH, born October 26, 1867.

70
REV. LUCIUS E. SMITH [11], son of Mary Torrey Smith (No. 62), born January 29, 1822; married Josephine F. Shattuck, of Groton, Mass., January 18, 1859.

Children:

CLARA J. A. SMITH, born February 5, 1860.

IDA S. P. SMITH, born September 23, 1861.

EDWARD C. SMITH, born July 26, 1869.

FLORA JOSEPHINE SMITH, born March 3, 1877.

71

ELISHA HUBBELL SMITH [11], son of Mary
Torrey Smith (No. 62), born April 13, 1825;
married Candace Marks, November 6, 1853; she
died August 3, 1871; married 2nd, Mrs. Anna
ELMS, September 11, 1877.

Children:

CARRIE MARILLA SMITH. born September 12,
1854; died April 18, 1883.
STEPHEN MARKS SMITH, born January 9, 1864.

———•••———

Descendants of Ephraim Torrey.

72

EPHRAIM TORREY [10], of Bethany, Pa., son of
William Torrey (No. 10), born in Williamstown,
Mass., March 26, 1782; married Eunice Lewis,
September 28, 1806; died September 24, 1829;
she died July 11, 1870, aged 80.

Children:

73 BETSEY TORREY, born November 6, 1808.
74 HANNAH WHEELER TORREY, born June 3, 1811.
75 MELISSA TORREY, born April 9, 1813.
LOIS TORREY. born July 7, 1815; died October
6, 1872.
76 CHARLOTTE S. TORREY, born February 21, 1817.
77 SETH L. TORREY, born January 1, 1819.
78 EMILY B. TORREY, born January 31, 1821.
79 OLIVE N. TORREY, born February 25, 1823.

80   WILLIAM WARD TORREY, born August 21, 1825.
81   ANN E. TORREY, born October 29, 1827.

73

BETSEY TORREY [11], daughter of Ephraim
(No. 72), born November 6, 1808; married Hiram
Devine, November 29, 1828; died April 19, 1847;
he died August 14, 1871.

Children:

EMMONS E. DEVINE, born October 1, 1829; married Rebecca A. Baney, October 12, 1874; children—*John B. Devine*, born September 23, 1875;
died September 17, 1877; *Charles T. Devine*,
born June 5, 1877; *Hiram E. Devine*, born March
26, 1879; *Samuel Devine*, born March 2, 1881.

EPHRAIM T. DEVINE, born May 4, 1831; died
April 24, 1859.

CHARLOTTE T. DEVINE, born March 24, 1833;
married John R. McKeen, October 28, 1860; he
died January 22, 1862; married 2nd, Edgar B.
Darling, September 19, 1872; he died June 11, 1883;
child—*Randolph W. McKeen*, born February 25,
1862.

LOWELL L. DEVINE, born June 8, 1835; died in
Autumn of 1837.

LUTHER W. DEVINE, born September 25, 1836;
married Emily J Ecker, January 2, 1864; children
—*Sherman N. Devine*, born September 27, 1864;
Incy L. Devine, born September 8, 1866; *Edith
J. Devine*, born July 15, 1868; *Florence R. Devine*,
born September 25, 1872; and *four other sons* who
died in infancy.

CELIA DEVINE, born March 9, 1839.
ALVIN DEVINE, born May 26, 1841.
MARY M. DEVINE, born May 14, 1843.
ALMA DEVINE, born March 4, 1845.

74

HANNAH W. TORREY [11], daughter of Ephraim (No. 72), born June 2, 1811; married Truman Goodnough, August 31, 1830.

Children:

LUCIAN H. GOODNOUGH, born June 16, 1833; married Margaret Pulis, October 6, 1857; she died March 20, 1860; married 2nd, Lucinda Starbird, November 16, 1870; children—*Chester J. Goodnough*, born October 10, 1871; *Frank V.* and *Vesta E. Goodnough* (twins), born November 5, 1873; *Eva P. Goodnough*, born January 14, 1876.

LOWELL GOODNOUGH, born August 13, 1838; married Martha Douglass, January —, 1871; she died June 15, 1883; children—*Mary Goodnough*, born September 27, 1871; *Benjamin Goodnough*, born August —, 1876.

LINAS GOODNOUGH, born January 3, 1841; married Alice G. Day, March 6, 1866; children—*Edwin D. Goodnough*, born January —, 1867; *Truman B. Goodnough*, born August —, 1869; *Grace Goodnough*, born January —, 1872; *Meroe Goodnough*, born January ——, 1875; *Gertrude Goodnough*, born January ——, 1878; *Luverne Goodnough*, born January —, 1881; *Jessie Goodnough*, born November —, 1883.

EMILY T. GOODNOUGH, born September 22, 1843; married James Davey, April —, 1869; children—*Lottie G. Davey*, born October 30, 1872; *Anna Davey*, born September 10, 1875; *Albina Davey*, born January —, 1883.

EUNICE GOODNOUGH, born June 10, 1847; died January 3, 1858.

ELIZA GOODNOUGH, born April 2, 1850; married John Boyd, September, —, 1868; he died March

1, 1876; married 2nd, Frank V. Carr, March 12, 1879; children—*Charles Boyd*, born July 27, 1872; *John Boyd*, born May 25, 1875; *Oscar Carr*, born December —, 1881.

CHARLOTTE GOODNOUGH, born April 13, 1853.

75

MELISSA TORREY [11], daughter of Ephraim (No. 72), born April 9, 1813; married John M. Horton, January 15, 1834; he died October 15, 1882.

Children:

CAROLINE S. HORTON, born May 1, 1835; married George S. Rice, December 21, 1862.

SEYMOUR C. HORTON, born March 11, 1837; married Lucy Jackson, November 24, 1872; children—*William M. Horton*, born December 7, 1873; *John S. Horton*, born September 18, 1876.

WILLIAM M. HORTON, born January 6, 1840; died in the Army, September 19, 1864.

RUTH E. HORTON, born November 22, 1842; married H. D. Wheeler, June 16, 1864; died December 5, 1882; children—*Edward M. Wheeler*, born September 4, 1865; *Ruth E. Wheeler*, born November 12, 1869; *Lucina Wheeler*, born October 7, 1871; *Arthur B. Wheeler*, born May 5, 1875; *Grace Melissa Wheeler*, born June 25, 1880.

EUPHEMIA E. HORTON, born September 28, 1844; married John A. Benedict, October 30, 1870.

EMILY L. HORTON, born February 7, 1849; died March 1, 1863.

76

CHARLOTTE S. TORREY [11], daughter of Ephraim (No. 72), born February 21, 1817; married

Enos Thatcher, February 3, 1843; he died August 26, 1882.

Children:

GEORGE W. THATCHER, born November 8, 1844; married Mary A. Fores, March 20, 1872; children—*Frank W. Thatcher*, born July 4, 1873; *Reno E. Thatcher*, born November 4, 1875; *Ada L. Thatcher*, born June 4, 1877; *Warren H. Thatcher*, born December 31, 1878; *Harry R. Thatcher*, born September 4, 1880; *Albert Thatcher*, born July 21, 1882.

CELIA H. THATCHER, born August 1, 1847; married John H. Allbee, March —, 1869; children—*George R. Allbee*, born November 20, 1869; *Marcia May Allbee*, born October 5, 1871; *Alice L. Allbee*, born July 21, 1875; died March 31, 1876; *Emma Ruth Allbee*, born August 25, 1883.

77

SETH L. TORREY [11], son of Ephraim (No. 72), born January 1, 1819; married Polly Alvira Wagner, May 24, 1848; she died August 14, 1851; married 2nd, Mary E. Butler, August 7, 1852.

Children by first wife:

WILLIAM E. TORREY, born September 4, 1849; died August 22, 1851.

MARY ALVIRA TORREY, born May 21, 1851; married Marvine Buckley————, 1875; he died November 9, 1882; children—*Twing R. Buckley*, born August 16, 1876; *Mabel F. Buckley*, born April 1, 1878; *Lewis M. Buckley*, born February 21, 1880.

MARTHA ANN TORREY, born May 21, 1851; died September 20, 1851.

Children by second wife:

MARTHA E. TORREY, born November 25, 1853;
married Clinton L. McOmber, November 25, 1874;
she died May 27, 1876.   No child.

ELSIE LOIS TORREY, born April 25, 1856.

LEWIS BUTLER TORREY, born October 2, 1862.

78

EMILY B. TORREY [11], daughter of Ephraim
(No. 72), born January 31, 1821; married John
M. Smith, March 14, 1844.

Children:

HENRY C. SMITH, born January 12, 1845; mar-
ried Louisa Berger, September 17, 1870; children
—*Victor H. Smith*, born July 3, 1871; *Freddie
Smith*, born July 8, 1872; died June 3, 1873; *Ed-
win G. Smith*, born August 24, 1873; *Kate I.
Smith*, born October 14, 1874; *Ira Smith*, born
January 25, 1876; died March —, 1876; *Walter
Smith*, born September 17, 1877; *Frank Smith*,
born February 8, 1881; *Kenneth Smith*, born
March 6, 1882; *Ida B. Smith*, born August 16,
1883.

MARY M. SMITH, born August 20. 1846; died
August 28, 1847.

WARREN S. SMITH, born December 6, 1847;
died January 6. 1852.

FRANKLIN S. SMITH, born October 27, 1849;
married Clara Taylor, June 9, 1872; children—
Clifford J. Smith, born April 14, 1874; *Elsie May
Smith*, born May 1, 1876; *Bessie Smith*, born July
—, 1878; *Eunice B. Smith*, born September 17,
1880; *Celia T. Smith*, born September 17, 1883.

George B. Smith, born April 27, 1851; married Emeline Bader, May 17, 1883; child— ——*Smith*, (a daughter) born April 12, 1884.

Filmore B. Smith, born April 29, 1853.

Emily T. Smith, born March 11, 1855; married Frank N. Dexter, May 16, 1883; child—*Martha T. Dexter*, born April 16, 1884.

Silas S. Smith, born August 28, 1857.

Howard J. Smith, born March 29, 1859.

Irving C. Smith, born December 1, 1860.

Ira J. Smith, born November 26, 1863; died March —, 1864.

79

OLIVE N. TORREY [11], daughter of Ephraim (No. 72), born February 25, 1823; married Josiah M. Bruce, in 1849.

Children:

Ward Bruce, born in 1850.

Robert Bruce, born in 1857.

The mother died in 1858.

80

WILLIAM WARD TORREY [11], (now of Greene, N. Y.,) son of Ephraim (No. 72), born August 21, 1825; married Lucrecia Tremain, July 4, 1849; she died May 5, 1871: married 2nd, Sarah Court, July 21, 1872.

Children:

Charles E. Torrey, born June 5, 1850; married Vesta A. Rogers, April 3, 1873.

WILLIAM E. TORREY, born March 22, 1856; married Flora E. Wright, December 24, 1881.

EVELYN L. TORREY, born March 30, 1858; married George H. Hinman, July 19, 1874; children —*Lucrecia E. Hinman*, born April 19, 1879; *Lillie B. Hinman*, born April 19, 1881.

Child by second wife:

MARY EUNICE TORREY, born October 25, 1876.

81

ANN E. TORREY [11], daughter of Ephraim (No. 72), born October 29, 1827; married David Saunders, in 1873; he died March 15, 1880; married 2nd, Joseph Stanton, in 1881.

———•••———

www.ingramcontent.com/pod-product-compliance
Lightning Source LLC
Chambersburg PA
CBHW022049210326
41519CB00055B/1196